AI or Not to AI

Authorship, Agency, and the Writer's Choice in a Human Led, AI Assisted World

R.A. Wilson

A Note to the Reader

This book is not an argument against technology. Nor is it in defense of it.

It is, instead, an attempt to clarify a question increasingly obscured by noise. What does it mean to write and to be responsible for writing, in an age of intelligent tools?

If you have picked up this book expecting a manifesto, a how-to guide, or a declaration of allegiance in a cultural conflict, you will not find that here. Those approaches tend to generate more heat than light. They presume conclusions before establishing terms. They frame disagreement as something to be defeated rather than understood.

That is not the work this book intends to do.

The discussion surrounding artificial intelligence and writing has accelerated rapidly, often outpacing careful thought. Claims are made with confidence but little precision. Fears are amplified. Assurances are overstated. And in the process, fundamental distinctions—between authorship and output, between agency and automation, between assistance and abdication—are frequently blurred or ignored altogether.

This book begins from the position that clarity matters.

It also begins from the position that writing has always involved tools. The quill, the press, the typewriter, the word processor, the search engine. Each altered how words were produced, revised, distributed, and evaluated. Each was met, in its time, with both enthusiasm and suspicion. And yet none of these tools, on their own, dissolved the idea of authorship. They changed the conditions of practice, not the responsibility of the writer.

Artificial intelligence introduces new questions, but not entirely new ones. What has changed is the speed with which text can be generated, the scale at which assistance is available, and the anxiety that accompanies those shifts. What has not changed is the fact that meaning does not arise from patterns alone. It arises from choice.

This book is concerned with those choices.

You will find no attempt here to “prove” that writing is human, nor to reassure readers through sentimentality. Nor will you find efforts to minimize the capabilities of contemporary AI systems. Such debates tend to circle endlessly, trading hypotheticals rather than addressing the realities of creative practice.

Instead, the focus throughout this book is deliberately narrower—and more demanding.

It asks where authorship actually resides.

It asks how agency is exercised, not assumed.

It asks why revision matters more than generation.

And it asks who remains responsible for words once they are released into the world.

These questions are not answered by detection software, nor by confidence scores, nor by surface-level assessments of style. They are answered through process, judgment, restraint, and accountability—qualities that do not appear fully formed in text, but are nonetheless essential to it.

Some readers may arrive at this book prepared to disagree. That is neither surprising nor unwelcome. Disagreement is often the sign of engagement. What this book does ask, however, is that disagreement be grounded in careful reading rather than reaction. The arguments that follow are not built on fear of technology, nor on faith in it. They are built on long-standing principles of craft and responsibility that predate the tools now under discussion—and that will almost certainly outlast them.

This is also not a book about evasion.

It does not offer strategies for concealment, nor tactics for bypassing scrutiny. Deception, whether human or automated, is a different problem entirely—and not one that can be solved by clever phrasing or technical workarounds. The question here is not how to avoid being examined, but how to understand what examination can and cannot reveal.

Authorship is not a property of text alone. It is a relationship between intention, action, and consequence. It involves decisions that occur before words are written and after they are revised. It entails responsibility for meaning, context, and impact. No tool—no matter how advanced—can assume that responsibility on behalf of a human being.

That responsibility remains where it has always been.

With the writer.

This book is written for those who care about writing as a practice rather than a product. For those who understand that clarity is not the enemy of creativity, and that restraint is often a form of respect—both for the reader and for the work itself. It is written for educators, editors, and thoughtful readers who sense that the current conversation has drifted away from first principles and are seeking a steadier footing.

You do not need to agree with every conclusion offered here. Agreement is not the goal. Understanding is. The chapters that follow are arranged not to persuade through force, but to proceed through clarification—each building upon the last, each grounded in the same central concern: how writers think, choose, revise, and ultimately stand behind their words.

If that question interests you, then this book invites you to continue.

If it does not, no argument here will compel you otherwise.

Either way, the conversation is worth having—carefully, calmly, and with due regard for what writing has always asked of those who practice it.

How to Use This Book

This book is not meant to be read once and set aside.

It is meant to be inhabited.

What follows is not a theory of artificial intelligence, nor a catalog of rules for acceptable use. It is an argument about authorship—where it actually lives, how it is exercised, and what it asks of those who claim it. That argument unfolds deliberately across the chapters, building from first principles rather than from policies or predictions.

For that reason, this book works in two distinct but complementary ways.

The first is conceptual.

The chapters are arranged to clarify a single question from multiple angles.
What does it mean to remain an author in a world where intelligent tools participate in the writing process?

Each chapter isolates one dimension of that question—fear, tools, agency, detection, revision, responsibility, and choice—not to provide a checklist of answers, but to restore distinctions that have become blurred. You are not expected to agree with every conclusion. You are expected to recognize where judgment must occur and where it cannot be delegated.

Read in this way, the book is an act of orientation.

It is meant to steady the conversation by returning it to fundamentals:
agency, revision, and responsibility.

The second way this book is meant to be used is practical.

The Appendix at the end distills the argument into a set of guiding principles for responsible, human-led, AI-assisted writing. It is not a code of conduct. It is not a compliance protocol. It is a scaffold for judgment.

That framework is designed to be used while writing, not merely reflected upon afterward.

You are invited to take a real piece of work—a draft, an assignment, an article, a chapter, a report—and move through the five principles deliberately:

to notice where generation occurred
to examine where revision reshaped meaning
to identify where refusal clarified intent
to ask where responsibility truly resides
and to decide what you are willing to stand behind

If you are a writer, you can use this framework to audit your own process—not to perform purity, but to clarify authorship.

If you are an educator, you can use it to shift assessment away from product and toward process—toward explanation, revision, and understanding.

If you are an editor or publisher, you can use it to frame disclosure and governance in terms of accountability rather than confession.

You may find it useful to return to the Appendix multiple times as you read.

Several chapters point toward it implicitly. Especially those on detection, revision, and responsibility. The framework is not an addendum to the argument. It is its operational form.

This book does not offer permission to use artificial intelligence.

It does not forbid it either.

What it offers instead is a posture:
that authorship is not proven by effort, nor dissolved by assistance, but enacted through judgment.

If you wish to test that posture, do so not by debating abstractions, but by applying it to your own work.

Ask, as you read:

Where am I governing language rather than merely producing it?

Where am I revising for meaning rather than plausibility?

Where am I refusing language that is fluent but untrue?

Where am I accepting responsibility rather than deferring it?

Those questions are not rhetorical.

They are the book.

Everything that follows is meant to clarify them.

The Map

Chapter One

The Question

The question is often posed badly.

Sometimes it is framed as a challenge—Is this writing real, or was it made by a machine? Other times it arrives disguised as a test—Can you prove you wrote this yourself? Increasingly, it appears as a warning—If machines can write, what happens to writers?

Each version carries its own assumptions, and nearly all of them misidentify the problem.

The issue is not whether artificial intelligence can produce text. It plainly can. Nor is the issue whether such text can be fluent, persuasive, or stylistically convincing. Anyone who has spent even a short time observing contemporary systems knows that this threshold has already been crossed.

The deeper question that refuses to stay settled, is something else entirely.

What does it mean to author a piece of writing when intelligent tools are part of the process?

This question is more difficult precisely because it resists easy answers. It cannot be resolved by pointing to a screen, a prompt, or a paragraph. It does not yield to detection software, nor does it disappear when technology improves. It sits beneath arguments about capability and performance, waiting to be addressed on its own terms.

To ask this question seriously is to recognize that writing has never been a purely mechanical act. Words have always been shaped by context, intention, revision, and consequence. Tools have always influenced how writing happens, but they have never fully accounted for why it happens—or who answers for it once it exists.

And yet, current discussions often flatten this distinction.

Much of the anxiety surrounding AI and writing comes from treating authorship as a visible property of text rather than an invisible process behind it. If writing looks a certain way, the reasoning goes, then it must have come from a certain kind of source. If a paragraph is polished, coherent, or stylistically consistent, it is sometimes assumed to be suspect—not because of what it says, but because of how easily it appears to say it.

This is a misunderstanding of both writing and tools.

Fluency is not authorship.
Coherence is not intention.
Pattern is not responsibility.

These qualities can be present in text generated by a system, but they are not sufficient to explain how meaning is formed or how judgment is exercised. Writing is not simply the arrangement of words into plausible order. It is the result of decisions—what to include, what to exclude, what to emphasize, what to revise, and what to stand behind once the work leaves the page.

The real question, then, is not whether AI can write.

It is whether writing can occur without agency.

And agency is not something a system possesses in the human sense. It does not choose to mean. It does not decide to be accountable. It does not experience consequence. These elements enter the process only when a human being directs, shapes, evaluates, and ultimately claims the work.

This is where confusion often enters the conversation.

When people argue that AI can be programmed to "fool" readers or systems, they are usually pointing to surface resemblance. They are describing imitation, not authorship. A convincing appearance can be produced without understanding; a persuasive rhythm can exist without responsibility. These facts are neither new nor unique to artificial intelligence. They have existed wherever technique has been mistaken for intention.

A skilled mimic can reproduce a voice.
A trained system can replicate a style.

Neither, by itself, answers the question of authorship.

That question remains anchored in who decides what the words are doing.

This distinction becomes especially important when writing is evaluated not only as expression, but as claim. Writing asserts ideas, frames reality, and influences others. It carries ethical weight precisely because it participates in shared meaning. When something is written and released, it enters a world where it can persuade, mislead, clarify, or harm. Responsibility for those effects cannot be delegated to a tool, no matter how sophisticated.

The presence of AI does not remove this burden.

It sharpens it.

As tools become more capable, the role of the writer becomes less about producing text and more about governing it. Judgment replaces effort as the central act. Selection matters more than speed. Revision matters more than generation. Restraint becomes a form of authorship in its own right.

This shift unsettles many familiar assumptions. For some, it feels like a loss of control. For others, it feels like an erosion of value. If writing becomes easier to produce, what distinguishes one writer from another? If assistance is available at scale, what remains uniquely human about the act?

These concerns are understandable. They arise whenever a craft confronts a tool that alters its visible labor. But they are also incomplete. They assume that what mattered most about writing was difficulty rather than discernment, effort rather than intention.

The question worth asking is not whether writing has become easier, but whether it has become less accountable.

And the answer to that question depends not on technology, but on how writers choose to engage with it.

This book proceeds from a simple premise. Authorship is not lost when tools are introduced, but it can be obscured when distinctions are ignored. The chapters that follow do not attempt to resolve every anxiety or anticipate every technological development. They return instead to fundamentals—agency, revision, responsibility, and choice—because those are the elements that determine whether writing remains human work.

The question, then, is not whether artificial intelligence belongs in the writing process. It already does.

The question is how writers understand their role once it is there.

That is the question at hand.

Chapter Two

The Fear

Fear is rarely precise.

It tends to arrive early, before language has caught up with experience, and it often attaches itself to the wrong object. What people fear is not always what is changing, but what they believe the change implies. In moments of technological transition, fear has a way of borrowing vocabulary from ethics while operating closer to identity.

The conversation around artificial intelligence and writing is no exception.

At first glance, the fear appears practical. Writers worry about displacement. Educators worry about integrity. Publishers worry about trust. Institutions worry about verification. Each of these concerns has a surface logic, and none of them should be dismissed out of hand. But taken together, they form a climate in which anxiety multiplies faster than understanding.

Beneath the stated concerns lies something quieter and more personal.

A fear of being unnecessary.

A fear of being exposed.

A fear that what once required effort can now be produced without it. And that the value of writing has been tied too closely to difficulty.

These are not technological fears.

They are existential ones.

The Fear of Replacement

Perhaps the most common anxiety is the fear that writers will be replaced.

The logic seems straightforward: if machines can generate text quickly and competently, then human writers become less valuable. If productivity increases without proportional effort, then the market will favor scale over craft. If words are no longer scarce, then those who produce them lose leverage.

This fear is understandable, but it rests on an incomplete understanding of what writing is valued for.

Writing has never been prized solely because it is difficult. Difficulty has often been incidental—a byproduct of limited tools, slower processes, and physical constraints. What has always mattered more is discernment: the ability to select what is worth saying, to shape it responsibly, and to stand behind it once it enters the world.

When writing is reduced to output alone, replacement becomes imaginable. When writing is understood as judgment, responsibility, and meaning-making, replacement becomes a category error.

Tools can scale production.
They cannot scale accountability.

That distinction matters more than speed.

The Fear of Exposure

Alongside replacement sits a subtler anxiety: the fear of being found out.

As AI-assisted writing becomes more visible, some writers worry not about losing work, but about losing legitimacy. If assistance is available, will others assume it was used? If a piece of writing is polished, will its quality itself become suspect? If effort is no longer visible, will authenticity be questioned?

This fear has already begun to shape behavior. Writers second-guess fluency. Educators scrutinize competence. Institutions search for signals that confirm their suspicions. In this environment, excellence risks being mistaken for automation, and confidence risks being read as concealment.

The underlying assumption is that authorship must now be proven defensively.

That assumption is corrosive.

Authorship has never been something that could be verified by surface inspection alone. It has always been inferred through consistency, process, and accountability over time. The sudden expectation that a paragraph should reveal its origins misunderstands how writing works and places an impossible burden on the text itself.

The fear of exposure, then, is not fear of wrongdoing.

It is fear of misinterpretation.

And misinterpretation thrives when trust erodes.

The Fear of Devaluation

There is also a broader fear—less personal, but no less potent—that writing itself is being devalued.

If anyone can generate a passable essay, a serviceable article, or a competent report, what distinguishes meaningful work from noise? If writing becomes abundant, does it become cheap? And if it becomes cheap, does it lose its power to shape thought?

This concern echoes earlier anxieties about mass printing, mechanical reproduction, and digital publishing. Each time, the volume of text increased. Each time, gatekeeping weakened. And each time, a familiar lament emerged: too many words, too little substance.

What history shows, however, is that abundance does not eliminate value—it relocates it.

When text is scarce, value attaches to access. When text is abundant, value attaches to judgment. Judgment is where authorship has always lived.

The skill that matters most is not the ability to produce words, but the ability to recognize which ones matter. This has always been true, but it becomes more visible when production accelerates.

The fear, then, is not that writing loses value.

It is that the criteria for value are changing—and that change exposes weak assumptions about where worth once resided.

The Fear of Cheating

In educational and institutional settings, fear often crystallizes around cheating.

If students can generate essays, if professionals can automate reports, if attribution becomes unclear, how can standards be maintained? How can learning be assessed? How can originality be protected?

These are serious concerns—but they are frequently misdirected.

Cheating is not defined by the use of tools.

It is defined by misrepresentation.

A student who submits work they do not understand, cannot defend, or did not meaningfully engage with has cheated, regardless of whether the words were typed by hand or assisted by software. Conversely, a student who uses tools transparently, thoughtfully, and in alignment with learning goals may be engaging more deeply than one who avoids assistance but memorizes patterns.

The fear of cheating often masks a deeper institutional discomfort: the realization that assessment methods have relied too heavily on product rather than process.

AI does not create this weakness.
It reveals it.

And revelation often feels like threat.

The Fear of Losing the Narrative

For writers in particular, there is another layer to the fear—one that is rarely stated outright.

It is the fear of losing the story we tell about ourselves.

Writers often identify not just with what they write, but with how they write. Effort, struggle, solitude, and persistence become part of identity. When tools alter the visible labor of writing, they disrupt this narrative. If the work becomes less arduous, does it become less authentic? If assistance is available, does choosing to use it feel like a betrayal of the self one has cultivated?

These questions are not trivial.

Identity is built on stories, and stories resist revision.

But the danger lies in confusing means with meaning.

The value of writing has never been found in suffering for its own sake. Difficulty can sharpen insight, but it can also obscure it. Tools that reduce friction do not automatically reduce depth. What matters is how that saved effort is reinvested—whether into thought, revision, and care, or simply into speed.

Fear arises when writers imagine that the presence of assistance negates intention.

It does not.

It merely shifts where intention must be exercised.

The Fear of Slippery Slopes

Another common anxiety is the fear of inevitability: the belief that once tools are introduced, agency will gradually erode until authorship disappears altogether.

This slippery-slope argument assumes that the use of assistance leads inexorably to abdication, that convenience will always override judgment, and that standards will inevitably collapse.

History offers little support for this view.

Tools do not force their use.

They invite it.

How they are adopted, regulated, and understood depends on cultural norms, institutional choices, and individual responsibility. Writing did not cease to be human when spelling was automated, when grammar was checked algorithmically, or when research became searchable. Each of these tools changed practice.

None of them eliminated authorship.

What the slippery-slope fear often reflects is not an analysis of tools, but a lack of trust—in systems, in people, or in oneself.

And mistrust, once established, seeks certainty where none is possible.

The Fear of Not Knowing Where the Line Is

Perhaps the most destabilizing fear of all is the sense that boundaries have become unclear.

Where does assistance end and authorship begin? How much help is too much? What counts as acceptable use, and who gets to decide?

These questions are real—and they are uncomfortable precisely because they resist universal answers. Context matters. Purpose matters. Transparency matters.

So does judgment.

The desire for bright lines is understandable. They simplify enforcement and relieve individuals of decision-making. But writing has never been governed by bright lines alone. It has always required interpretation, discretion, and ethical consideration.

AI does not eliminate these requirements.

It intensifies them.

Fear arises when people mistake ambiguity for absence—when the lack of simple rules is taken to mean that no standards exist at all. In reality, standards remain.

They simply demand more thought.

Naming the Fear Clearly

What unites these fears is not technology, but uncertainty about value, identity, and responsibility.

Writers fear being replaced because they equate value with effort.

They fear exposure because they equate legitimacy with

visible labor.
Institutions fear cheating because they equate learning with output.
Readers fear devaluation because they equate abundance with emptiness.

Each fear contains a partial truth.

Each becomes misleading when taken as a whole.

The presence of AI does not remove the need for writers. It removes the illusion that writing was ever only about typing words into place. It forces a confrontation with what has always mattered but was easier to ignore when labor was slow and tools were blunt.

Fear thrives when distinctions collapse. Clarity begins when they are restored.

The chapters that follow do not attempt to soothe fear by minimizing change, nor to inflame it by exaggerating consequences. They proceed instead by separating what is genuinely new from what is newly visible—by distinguishing between threats to craft and challenges to habit.

Fear is not the enemy of this conversation.

Confusion is.

And confusion yields only when the terms of the discussion are carefully, patiently reclaimed.

That work begins by understanding the tool itself.

Chapter Three

The Tool

Tools become most confusing when they are treated as actors.

In discussions about artificial intelligence and writing, language often slips—subtly but persistently into forms that assign agency where none exists. Systems are described as deciding, knowing, understanding, or intending. They are said to want, to learn, to replace, or to compete. These verbs do not merely describe function; they imply personhood.

This rhetorical drift is not accidental. It reflects a long-standing habit of human thought: when confronted with complex mechanisms, we anthropomorphize them. We do so not because the tools demand it, but because our language struggles to describe systems that act without consciousness and produce results without understanding.

Clarity begins by resisting that impulse.

Artificial intelligence, as it is used in writing today, is not an author, a collaborator, or an independent agent. It is a tool—one that operates by identifying and extending patterns in language based on statistical relationships. It does not possess intention. It does not experience meaning. It does not hold beliefs, values, or goals of its own.

What it produces is output, not authorship.

What the Tool Actually Does

At its most basic level, contemporary AI writing systems analyze large bodies of text and learn to predict which words are likely to follow others in given contexts. When prompted, they generate sequences of language that reflect those probabilities. The result can appear fluent, coherent, and even persuasive—but appearance should not be mistaken for comprehension.

The system does not know what it is saying—indeed, that it is saying anything at all.

It does not evaluate truth, relevance, or consequence in the human sense. It does not care whether a sentence is misleading, profound, or harmful. It cannot distinguish between insight and cliché, between wisdom and error, except insofar as those distinctions are reflected indirectly in patterns it has already absorbed.

This is not a flaw. It is the nature of the tool.

Confusion arises when people expect more—or fear more—than the tool is capable of being.

The Illusion of Understanding

Because AI-generated text can be smooth and contextually appropriate, it often gives the impression of understanding. This impression is reinforced by conversational interfaces that mirror human exchange. Questions are asked. Answers are returned. The rhythm feels familiar.

But familiarity does not equal cognition.

The system does not understand a question in the way a human does. It does not grasp intention, nuance, or implication. It responds by mapping the prompt onto patterns it has learned and extending those patterns forward. The result can be useful—sometimes remarkably so—but usefulness is not the same as comprehension.

This distinction matters because understanding is the foundation of responsibility.

A tool that does not understand cannot be accountable. It cannot explain why it produced one response instead of another. It cannot defend a claim. It cannot revise based on moral judgment. All of these actions require an interior perspective the tool does not possess.

Responsibility therefore remains outside the system.

It resides with the human who chooses to use it, interprets its output, revises its suggestions, and decides what enters the world under their name.

Assistance Versus Delegation

One of the most persistent confusions in the AI debate is the failure to distinguish between assistance and delegation.

Assistance supports a process while leaving control intact.

Delegation transfers responsibility to another agent.

AI can assist writing. It cannot meaningfully accept delegated responsibility for it.

When a writer uses a tool to brainstorm, outline, rephrase, or explore alternatives, the writer remains the decision-maker. They evaluate relevance. They accept or reject suggestions. They shape the final form. The tool provides possibilities; the human provides judgment.

Problems arise only when this distinction is ignored—when output is accepted uncritically, when revision is abandoned, or when authorship is treated as something that can be outsourced entirely. In such cases, the failure is not technological. It is human.

The tool does not insist on trust, demand authority, or claim authorship.

Those choices are made elsewhere.

Why the Tool Feels Different

Despite these continuities, AI feels different from earlier writing tools—and that difference should be acknowledged honestly.

Unlike spellcheckers or grammar engines, AI can generate extended passages of text. It can propose structure, tone, and argument. It can operate at scales and speeds that outstrip previous forms of assistance. This amplifies both its usefulness and its risks.

But amplification is not transformation.

The presence of generation does not convert a tool into an author any more than a camera converts itself into an artist. The camera can capture an image. It cannot decide what the image should mean. That decision remains with the photographer.

What AI changes is the visibility of decision-making. When words appear quickly, the labor of producing them recedes. What remains exposed is the labor of selection, revision, and responsibility. Writers who rely on those skills feel the shift less acutely. Writers who equated authorship with effort feel it more.

The tool reveals where the work truly was.

Pattern Without Purpose

It is worth emphasizing what AI systems lack. Not to diminish them, but to place them correctly.

They do not possess goals, beliefs, values, or intentions. They have no awareness of audience and no concern for consequence.

They cannot weigh competing goods. They cannot recognize irony except where it appears statistically. They do not know when silence is better than speech. They cannot choose restraint for ethical reasons.

They generate language, not meaning. Meaning arises when a human being:

selects one phrasing over another
shapes emphasis
withholds what should not be said
revises in response to context
accepts responsibility for how words will be received

These acts do not emerge from pattern prediction. They emerge from judgment.

The Tool as Mirror

One of the least discussed aspects of AI-assisted writing is the way it reflects the user.

The quality, coherence, and integrity of output are often correlated with the clarity of the prompt, the discernment of the editor, and the care of the reviser. Vague thinking produces vague results. Careless acceptance produces shallow prose. Thoughtful engagement produces something better.

In this sense, AI functions less as an author and more as a mirror—amplifying strengths and weaknesses already present in the user.

This is unsettling to some, because it removes a convenient scapegoat. If the writing lacks depth, the problem cannot be blamed entirely on the tool. If the argument lacks coherence, no system can supply it whole. The tool can extend patterns; it cannot replace thought.

The discomfort many feel is not about loss of control, but exposure of responsibility.

Consider a researcher drafting the introduction to an article. They prompt the system with a vague request and receive a vague paragraph in return. Grammatically smooth, but conceptually thin.

Unsatisfied, they revise the prompt, clarifying the argument and the audience. The next response is sharper, but still misplaces the central claim.

The researcher edits the paragraph, cuts a sentence that overstates the evidence, and adds a qualifying phrase. The final version contains fragments of the system's suggestion, but the argument now reflects the researcher's judgment. The authorship lives in those decisions: what was kept, what was removed, and what was rewritten to carry the intended meaning.

What the Tool Cannot Do

It is equally important to be clear about what AI cannot do, regardless of future improvements in fluency.

It cannot decide why something should be written.

It cannot determine whether something ought to be written at all.

It cannot weigh harm against benefit.

It cannot stand behind a claim when challenged.

These are not technical limitations awaiting resolution. They are categorical boundaries. No increase in data, speed, or scale converts a system without consciousness into one that bears responsibility.

When these boundaries are ignored, expectations become distorted. AI is alternately feared as an unstoppable agent or dismissed as a trivial toy. Both positions miss the point.

The tool is powerful because it accelerates processes humans already perform.
It is dangerous only when those humans abandon judgment.

Reclaiming Proportion

Much of the current debate about AI and writing suffers from a lack of proportion.

Capabilities are overstated. Risks are inflated. Futures are predicted with unwarranted certainty. In this atmosphere, the tool becomes a symbol rather than an object of analysis—standing in for broader anxieties about work, value, and change.

Reclaiming proportion means returning to fundamentals.

Writing is not defined by how text appears on a page. It is defined by how meaning is shaped, revised, and owned. Tools may alter the mechanics of production, but they do not alter the moral architecture of authorship.

The presence of AI does not absolve writers of responsibility.

It concentrates it.

As assistance increases, so does the importance of discernment. As generation accelerates, so does the need for restraint. The tool does not remove the writer from the process; it demands more from them.

The Tool in Its Proper Place

When placed correctly, AI becomes neither savior nor threat.

It becomes what all tools have always been: an extension of human intention, constrained by human judgment, and answerable to human responsibility. It can support exploration. It can accelerate drafting. It can surface alternatives. But it cannot decide what matters.

That decision belongs elsewhere.

Understanding the tool clearly is not the end of the conversation. It is the condition for continuing it responsibly. Once the nature of the tool is understood, the focus can shift to the concept it so often obscures.

Authorship.

Chapter Four

Authorship

Authorship is easy to invoke and difficult to define.

The word carries cultural authority. It appears on title pages, legal documents, and institutional policies. It is associated with originality, ownership, and responsibility. And yet, when examined closely, authorship resists simple boundaries. It is not a property that can be isolated in a sentence or proven through inspection. It is a role—a relationship between a human being and a body of work that unfolds over time.

This is where debates about artificial intelligence so often falter.

When authorship is misunderstood as a visible feature of text rather than an invisible process behind it, confusion follows. Fluency becomes suspect. Assistance becomes conflated with abdication. And writing is treated as though it must reveal its origins in its surface characteristics.

It cannot.

Authorship does not reside in phrasing alone. It resides in agency.

Authorship as Agency

To author something is not merely to cause it to exist. It is to decide what it will be, why it will exist, and what one is willing to answer for once it does.

Agency, in this sense, is not a moment. It is a sequence of decisions.

Deciding to write at all

What is worth saying and how it should be framed

What to exclude

When it is finished and whether one is willing to stand behind it publicly

None of these decisions are contained in text. They surround it.

A tool can contribute language.
It cannot originate purpose.

This is the boundary that defines authorship—and it remains intact regardless of how advanced the tool becomes.

Why Text Alone Cannot Prove Authorship

There is a persistent belief that authorship can be detected in style, complexity, or imperfection. According to this view, human writing is assumed to bear certain markers—idiosyncrasies, inconsistencies, or inefficiencies—that distinguish it from machine-generated text.

This belief has never been reliable.

Human writing has always ranged from rough to polished, from idiosyncratic to formulaic. Entire literary movements have been built on imitation, constraint, and formal rigor. Conversely, machines are now capable of producing text that appears spontaneous, uneven, or idiosyncratic—because those patterns, too, exist in the data they model.

Surface features are not evidence of authorship.

They are artifacts of style.

Authorship is not proven by how something looks. It is inferred through control, coherence over time, and responsibility for consequence. These qualities emerge only when a human mind governs the work across its full lifecycle.

Intention and Meaning

Meaning does not arise automatically from words.

It arises from intention interacting with context.

A sentence can be grammatically correct and entirely meaningless. It can be fluent and misleading. It can be elegant and irresponsible. What gives writing weight is not correctness, but deliberateness—the sense that someone chose these words knowing they would be read, interpreted, and acted upon.

This is where authorship becomes inseparable from ethics.

An author is not simply someone who produces text, but someone who accepts that words do things in the world. They persuade. They frame reality. They shape belief. To author is to accept that influence knowingly.

A system that generates language cannot do this. It cannot intend meaning, weigh consequence, or accept blame or credit.

Those responsibilities do not vanish when tools are introduced. They remain human obligations.

Revision: Where Authorship Becomes Visible

If authorship is difficult to locate in first drafts, it becomes clearer in revision.

Revision is not merely correction. It is interpretation. It is where a writer confronts what has been produced and decides what it should become. It involves judgment, restraint, and often refusal—the willingness to discard language that is fluent but wrong, persuasive but misleading, or clever but empty.

Revision is where authorship asserts itself most clearly.

A writer revises because they care about meaning. They revise because they anticipate a reader. They revise because they recognize responsibility. These motivations do not arise from pattern prediction. They arise from awareness.

AI systems do not revise in this sense. They regenerate. They offer alternatives. But they do not choose among them with moral or contextual awareness. That choice remains external to the tool.

Where revision is absent, authorship thins.
Where revision is deliberate, authorship deepens.

Originality Reconsidered

One of the most emotionally charged aspects of authorship is originality.

Originality is often imagined as creation from nothing—a solitary act of invention untainted by influence. This image has always been a myth. Writers learn by reading. They absorb forms, rhythms, and conventions. They borrow, adapt, and transform what came before.

Originality has never meant isolation.

It has meant selection and synthesis.

AI-assisted writing makes this process more visible. When a tool proposes language based on existing patterns, it exposes what was always true: that writing emerges from engagement with prior texts. What distinguishes authorship is not the absence of influence, but the presence of judgment.

The author decides:

which influences matter
which patterns are appropriate
which formulations serve the purpose at hand

The tool does not.

Originality survives not because humans avoid assistance, but because they exercise discernment.

Ownership and Responsibility

Authorship is inseparable from ownership—not merely legal ownership, but moral responsibility.

To author a work is to say, implicitly or explicitly: these words represent me. This claim carries weight precisely because it binds the writer to the effects of the work. Praise and criticism alike assume that someone chose to put these words into circulation.

AI cannot make that claim.

When responsibility is obscured—when authorship is treated as ambiguous or transferable—trust erodes. Readers do not need to know every tool used in the creation of a text, but they do need to know that someone stands behind it. Someone who can answer questions. Someone who can revise or retract. Someone who can be held accountable.

This is why authorship matters more, not less, in an AI-assisted world.

The Error of Treating Authorship as a Threshold

Many institutional responses to AI rely on thresholds: how much assistance is too much? Where is the line between help and authorship? At what point does a writer cease to be the author?

These questions assume authorship is a quantity. It is not.

Authorship is not measured by percentage of text generated or hours spent drafting. It is measured by control. A writer may accept extensive assistance and still exercise full authorship if they govern purpose, meaning, revision, and release. Conversely, a writer may type every word themselves and still abdicate authorship if they act without understanding, intention, or responsibility.

This is uncomfortable because it resists automation.

But writing has always resisted automation at the level that matters.

Authorship in Practice

In practical terms, authorship manifests through patterns of behavior rather than properties of text:

the ability to explain why choices were made
the willingness to revise when meaning fails
the capacity to defend or reconsider claims
the consistency of voice and judgment over time

These are not detectable through snapshots. They emerge through engagement.

This is why serious readers, editors, and educators rely on process, dialogue, and sustained interaction—not on isolated artifacts. Authorship reveals itself across drafts, conversations, and revisions. It cannot be reverse-engineered from a paragraph.

Why Authorship Still Matters

Some argue that authorship itself is becoming obsolete—that in an age of abundant text, attribution matters less than utility. If information is accurate and useful, why concern ourselves with who produced it?

This argument misunderstands the role writing plays in human culture.

Writing is not only a means of transmitting information. It is a means of shaping shared understanding. When authorship dissolves, responsibility disperses. When responsibility disperses, trust collapses.

Authorship is not a sentimental relic.

It is a structural necessity.

Without it, there is no one to question, no one to correct, no one to answer for harm or error. The presence of tools does not eliminate this need. It intensifies it.

Reclaiming Authorship

Reclaiming authorship in an AI-assisted world does not require rejecting tools. It requires placing them correctly.

The writer remains the one who:

sets intent
governs meaning
revises with care
accepts responsibility

The tool assists where invited and stops where directed.

This relationship is not fragile. It is robust—provided distinctions are maintained and agency is not surrendered for convenience.

Authorship survives not because machines fail, but because humans choose to remain accountable.

That choice is the foundation of writing as a human practice.

And it is the foundation on which the rest of this book rests.

Authorship as a Lived Claim

Authorship, as this book has argued, is not a credential conferred by process nor a status inferred from effort. It is a claim made through judgment. To author a text is to say—not merely that one produced it, but that one stands behind what it means and what it does in the world.

This posture does not depend on how language entered the page. It depends on what happened after. Authorship becomes visible where a writer chose one phrasing over another, where an argument was narrowed rather than expanded, where a fluent sentence was removed because it misrepresented what the writer actually believed. These moments rarely announce themselves. They leave no forensic trace. And yet they are the moments in which authorship quietly asserts itself.

To treat authorship as a lived claim rather than a technical property is to shift attention away from tools and toward responsibility. It is to recognize that what ultimately matters is not whether assistance was used, but whether judgment was exercised, whether revision was undertaken in good faith, and whether the writer remains willing to answer for the meaning that now circulates under their name.

For writers who wish to test that posture, a few quiet questions are often more revealing than any declaration:

What am I actually trying to say here?
What would it mean for me to stand behind this version?
Where did judgment, not fluency, change this text?

These questions do not certify authorship. They enact it.

If authorship is a lived claim, detection can never prove it. Only practice can.

Chapter Five

Detection & Deception

When uncertainty rises, institutions reach for instruments.

In the current debate over artificial intelligence and writing, detection software has become one such instrument—promoted as a means of restoring certainty, preserving integrity, and distinguishing human authorship from automated output. Its appeal is obvious. In a climate of mistrust, detection promises resolution. It offers a way to name what feels ambiguous, to draw a line where none is easily visible.

But the confidence placed in detection often exceeds what detection can deliver.

This gap, between expectation and capability, is where much of the conversation goes wrong, and it becomes clearer once we examine how detection systems actually operate.

The Desire to Detect

The impulse to detect AI-generated writing is not rooted primarily in technology.

It is rooted in fear.

Institutions worry about erosion of standards. Educators worry about learning outcomes. Publishers worry about credibility. Each of these concerns is legitimate, but detection is often treated as a universal remedy rather than what it actually is: a limited, probabilistic signal operating under narrow assumptions.

Detection software does not identify authorship. It estimates likelihood.

That distinction matters.

Most detection systems rely on machine-learning models trained to recognize statistical patterns associated with machine-generated text. In other words, artificial intelligence is often used to detect artificial intelligence.

These systems are trained on large datasets containing both human and AI-produced writing, and they learn to classify new text based on probability distributions. How predictable words are, how sentences vary, and how patterns appear across a passage. The result is not a determination of authorship but a statistical guess about resemblance.

When detection systems analyze text, they are not observing the writing process. They are examining the surface patterns of the final language and asking whether those patterns resemble examples in their training data.

Authorship is a human relationship to a text—one that unfolds through intent, revision, and responsibility. Detection tools do not have access to these dimensions. They analyze surface features: statistical regularities, token distributions, sentence entropy, and stylistic predictability. From these indicators they infer estimated probability, not proof.

How Detection Actually Works

AI-detection systems are trained to recognize patterns commonly associated with machine-generated text. These patterns are not intrinsic to AI, but incidental to how certain models tend to produce language when operating without human intervention or revision.

As a result, detection performs best on:

First-pass outputs
Unedited generation
Generic prompts
Short-form text

It performs worst on:

Revised work
Collaborative drafting
Long-form writing
Texts shaped by human judgment

This is not a flaw to be fixed.

It is a structural limitation.

Once a human meaningfully edits a text—revises structure, alters emphasis, reorders arguments, refines tone—the statistical signals detection relies on dissolve. At that point, the text reflects human decision-making layered atop assistance. There is no technical method to reverse that process.

Detection cannot see revision, and revision is where authorship lives.

The Myth of Forensic Certainty

Despite these limitations, detection tools are often described in forensic terms. They are imagined as objective arbiters capable of uncovering hidden origins, exposing deception, and rendering final judgment.

This framing is deeply misleading.

There is no reliable way to determine authorship from text alone. This has been true long before artificial intelligence entered the conversation. Stylometry, authorship attribution, and plagiarism detection have always operated probabilistically, with margins of error that increase dramatically as texts grow longer and more complex.

AI does not change this reality.

It intensifies it.

Detection scores—percentages, confidence levels, risk ratings—create the impression of precision. But precision is not accuracy. Numbers can obscure uncertainty as easily as they can clarify it.

A high score does not prove deception, and a low score does not prove authorship.

What detection offers is not truth, but a guess framed in technical language.

Academic Example:

The accusation arrived without a voice.

It appeared first as a notation in a system the writer did not control, then as a brief message requesting clarification. No passage was cited. No claim was identified as false or misunderstood. Only a classification appeared—likely assisted—followed by a procedural pause.

The writer reread the work, searching for whatever might have triggered suspicion. The sentences were spare. The argument was narrow. Nothing felt borrowed or inflated. The writing bore the marks of constraint rather than excess—choices made by omission as much as inclusion.

When asked to explain the process, the writer did so carefully. Drafts had been written, discarded, rewritten. Language that felt persuasive but inaccurate had been removed. Several paragraphs had been rewritten late in the process, not to improve fluency, but to prevent overstatement.

The explanation was noted.

It was not weighed.

The reviewer did not claim deception. They did not need to. The system's output was treated as sufficient cause for doubt. Not proof—just enough uncertainty to shift the burden. The writer was now asked to demonstrate authorship, as though authorship were something that could be produced on demand.

No further evidence was requested.

No counter-evidence was considered.

The process moved forward without resolution, leaving the work intact but the writer marked—not guilty, not innocent, simply suspect.

What had changed was not the text.

It was the presumption beneath it.

Deception as a Misplaced Focus

Much of the rhetoric surrounding AI and writing assumes deception as a central threat. Writers are imagined as attempting to “fool” systems. Students are presumed to be seeking concealment. The debate becomes adversarial by default.

This framing misidentifies the problem.

Deception is not primarily a technological issue.

It is an ethical one.

A person who intends to deceive can do so with or without AI. Conversely, a person who acts transparently and responsibly does not become deceptive simply because they use tools.

The presence of AI does not create dishonesty.

It exposes it.

Focusing on deception shifts attention away from the real question: how do we evaluate authorship and learning in a world where assistance is ubiquitous? Detection does not answer that question.

It postpones it.

The False Binary of "Human" vs. "AI"

Detection tools rely on a binary that does not reflect reality: text is classified as either human-written or AI-generated. But most contemporary writing exists somewhere in between.

Writers brainstorm with tools. They rephrase. They test alternatives. They revise heavily. They discard most of what is suggested. The final work reflects human judgment layered over assistance—not pure generation from either source.

Detection systems are not designed to recognize this hybridity.

As a result, they produce false positives—flagging human-authored work as suspect—and false negatives—missing uncritical AI output entirely. Both outcomes undermine trust.

When writers are penalized for fluency, clarity, or coherence, standards invert. Excellence becomes suspicious. Care becomes risky. The system begins to punish the very qualities it claims to protect.

This is not integrity.

It is distortion.

The Institutional Shortcut

Why, then, do institutions continue to rely on detection?

Because detection offers a shortcut.

Process-based evaluation is difficult. It requires dialogue, drafts, explanation, and judgment. Detection offers an automated alternative. Scalability without engagement.

But shortcuts come with costs.

When institutions outsource judgment to tools, they tacitly concede that authorship can be inferred mechanically. This concession undermines the very standards they seek to uphold. It treats writing as an artifact rather than an act, a product rather than a practice.

AI detection does not fail because it is poorly designed.

It fails because it is asked to answer the wrong question.

The Danger of Overconfidence

The most serious risk posed by detection tools is not their inaccuracy, but the confidence placed in them.

When a detection score is treated as evidence rather than signal, harm follows. Writers are accused without proof. Students are disciplined without understanding. Trust erodes—not because of AI, but because authority has been misapplied.

The irony is difficult to ignore.

Tools introduced to preserve integrity often undermine it.

This is not because technology is malicious, but because it is misunderstood. Detection systems do not know when they are wrong. They cannot explain their conclusions.

They cannot contextualize their results. They cannot account for revision, collaboration, or intent.

They should never be treated as arbiters.

Deception Versus Disclosure

There is an important distinction between deception and disclosure.

Disclosure is not confession.

It is contextual clarity.

It acknowledges that tools were used without surrendering authorship or responsibility.

Many publishers and institutions are already moving toward language that recognizes AI-assisted drafting while affirming human-led authorship.

This shift reflects a growing understanding: that transparency matters more than purity, and that honesty about process is more sustainable than attempts at concealment.

Detection cannot enforce honesty.

Only culture can.

The Futility of the Arms Race

Some argue that detection will improve, that future systems will be able to identify AI involvement with greater certainty. This belief fuels an arms race mentality: tools to generate, tools to detect, tools to evade, tools to counter-evade.

This race is unwinnable.

As long as humans revise, collaborate, and exercise judgment, there will be no stable signature to detect. As long as writing involves thought rather than transcription, authorship will remain irreducible to statistics.

The pursuit of perfect detection misunderstands the nature of writing.

It is chasing a certainty that does not exist.

What Detection Reveals Instead

Ironically, the rise of detection tools has revealed something valuable—not about AI, but about institutions.

It has exposed how much evaluation relied on surface cues rather than engagement. It has shown where trust was assumed rather than built. And it has highlighted the need for clearer definitions of authorship, learning, and responsibility.

Detection does not solve these problems.

It points toward them.

Reframing the Conversation

The question is not whether AI can fool systems.

The question is whether systems can adapt to a reality where authorship must be understood through process, not appearance.

This requires:

Valuing revision over output.
Prioritizing explanation over inspection.
Trusting judgment over automation.

These shifts are uncomfortable because they resist scale. They require time, dialogue, and human discernment. But they are not optional. Writing has always demanded them, even when institutions could afford to ignore that fact.

AI removes that luxury.

The End of the Illusion

Detection and deception dominate the conversation because they offer the illusion of control. They promise a return to certainty through technical means. But certainty about authorship has never been technical.

It has always been relational.

Authorship exists where someone can say, without hesitation: I chose these words, I revised them, and I stand behind what they do in the world.

No system can replicate that claim.

No detector can verify it.

And no amount of technological sophistication will change the fact that writing remains, at its core, a human responsibility.

The failure of detection is not a crisis.

It is a reminder.

Chapter Six

Revision

If authorship has a visible signature, it is not found in generation.
It is found in revision.

Revision is where writing ceases to be provisional and becomes deliberate. It is the moment when language stops being merely possible and begins to be accountable. Every serious writer knows this instinctively, even if they do not always articulate it. The first draft explores. The second interrogates. The third decides.

This has always been true.

Artificial intelligence does not change the centrality of revision.

It clarifies it.

Revision as Judgment, Not Correction

Revision is often misunderstood as cleanup—a mechanical process of fixing errors, smoothing transitions, and correcting grammar. That work matters, but it is not what defines revision as an authorial act.

True revision is interpretive.

It asks questions no tool can ask on its own:

Is this what I mean?

Does this framing mislead?

What am I implying by saying this now, this way?

What must be removed for the meaning to hold?

These questions are not technical. They are ethical, rhetorical, and contextual. They require awareness of audience, consequence, and intent.

A tool can suggest alternatives.
It cannot decide which alternative should exist.

That decision belongs to the author.

The Moment of Refusal

One of the clearest signals of authorship is refusal.

Writers revise not only by adding, but by discarding. They remove passages that are fluent but false, persuasive but irresponsible, clever but hollow. They learn—often painfully—that just because language works does not mean it belongs.

This act of refusal is invisible to detection systems. It leaves no statistical trace. And yet it is one of the most human moments in the entire writing process.

AI systems do not refuse language.

They generate until stopped.

They do not feel the tension between saying something well and saying something right. They do not experience the discomfort of realizing that a sentence is eloquent but wrong. They do not hesitate.

Writers do.

And that hesitation—the pause before deciding whether language should remain—is where authorship asserts itself most clearly.

Revision as Responsibility

Revision is also where responsibility enters the process explicitly.

A writer revises because they anticipate being read. They imagine misunderstanding. They foresee objections. They recognize harm. They refine not only to improve clarity, but to limit damage.

This anticipatory responsibility distinguishes writing from mere transcription.

When a writer revises a claim, they are not optimizing output. They are negotiating meaning in a social world. They are asking how words will function once detached from their original context and released to readers who bring their own assumptions, histories, and vulnerabilities.

AI systems do not anticipate responsibility.

They cannot imagine consequences beyond pattern continuation. They cannot feel the weight of being misinterpreted. They cannot regret.

Revision, therefore, is not simply craft.

It is ethics in practice.

The Illusion of Instant Writing

One of the reasons AI unsettles writers is that it collapses the visible timeline of drafting. Text appears quickly, creating the impression that writing itself has been trivialized. The struggle that once accompanied composition is no longer visible.

But the struggle was never the essence of writing. It was a byproduct of constraints.

Revision reveals this truth. Even when generation is fast, revision remains slow. Judgment cannot be accelerated beyond a certain point. Meaning cannot be verified at machine speed. The more language is produced, the more discernment is required to shape it responsibly.

Speed shifts the focus of effort.

It does not eliminate it.

Writers who understand this feel less threatened by AI, because they recognize that the work they value was never in typing.

It was in deciding.

Revision as Interpretation

Every act of revision is an interpretation of what already exists.

When a writer revises, they are not merely altering text. They are interpreting their own work—deciding what it says, what it implies, and what it must become. This recursive relationship between writer and text is central to authorship.

The writer becomes the reader of their own work.

They ask not only Is this effective, but Is this faithful to my intent? They discover meanings they did not consciously place there. They recognize implications they did not foresee. Revision becomes a dialogue between intention and outcome.

AI systems cannot participate in this dialogue.

They can propose variants, but they do not interpret meaning. They do not recognize unintended implication. They do not experience surprise at what language has done.

That surprise—the moment when a writer realizes what their words actually say—is deeply human.

Revision Across Time

Authorship also reveals itself through revision over time.

Writers return to work with distance. They revise with memory, experience, and altered perspective. What once seemed essential now appears indulgent. What once felt complete now feels thin.

This temporal dimension of revision is inseparable from authorship.

AI systems do not grow in this way.

They update models, but they do not develop perspective. They do not revise because they have changed. They revise only when prompted.

Human writers revise because they are not the same person they were when the words first appeared.

That difference matters.

Revision and Accountability

When challenged, authors respond through revision. They clarify, correct, retract, and refine.

These acts assume responsibility not only for what was written, but for how it was received. They recognize that writing participates in a shared world where meaning is negotiated rather than fixed.

AI cannot do this.

It cannot be held to account. It cannot explain its choices. It cannot acknowledge error as error rather than as deviation from pattern. Any revision it produces in response to criticism is still governed by probability, not accountability.

The presence of revision, then, is not merely evidence of human involvement.

It is evidence of authorship.

Why Revision Confounds Detection

Detection systems struggle precisely where revision is most active.

Heavily revised work bears fewer statistical markers. It reflects idiosyncratic decisions layered over suggestions. It contains contradictions resolved through judgment rather than probability. It reflects restraint rather than optimization.

Not because revision is designed to evade detection.

It is because revision is where writing stops behaving like data.

The more a human revises, the less the text resembles the average patterns detection systems rely on. The text becomes specific—tied to a mind, a context, a purpose. Detection cannot isolate that specificity because it is not a surface feature.

It is a relationship.

Revision as the Answer to the Fear

Much of the fear surrounding AI-assisted writing dissolves when revision is restored to its central place.

The fear of replacement weakens when writers recognize that judgment cannot be automated.

The fear of exposure weakens when writers understand that authorship is not proven by effort but by responsibility.

The fear of devaluation weakens when writing is understood as meaning-making rather than production.

Revision makes these distinctions visible.

It reminds writers that what they contribute is not speed, but discernment. Not volume, but care. Not fluency, but fidelity to meaning.

Teaching Revision, Not Policing Generation

In educational contexts, revision offers a path forward that detection cannot.

Rather than policing generation, educators can emphasize revision—requiring students to explain choices, reflect on drafts, and demonstrate understanding. These practices assess learning directly rather than inferring it indirectly through suspicion.

AI does not undermine this approach.

It reinforces it.

Students who cannot revise meaningfully cannot hide behind generated text for long. Students who can revise demonstrate authorship regardless of how language entered the page initially.

Revision shifts assessment from product to process.

Where it belongs.

Revision as the Writer's Claim

At the end of the process, revision becomes the writer's claim to the work.

To revise is to say: This is what I mean
To revise is to say: I have considered alternatives
To revise is to say: I accept responsibility for this version

That claim cannot be made by a tool.

It can only be made by a human being willing to stand behind their words.

The Center Holds

If authorship feels threatened in an age of intelligent tools, it is not because writing has lost its human core.

It is because that core has been misunderstood.

Revision reveals what has always been true: writing is not the act of producing language, but the act of governing it. Tools may assist in production. They cannot replace governance.

As long as writers revise with judgment, care, and accountability, authorship remains intact.

Not hidden.
Not diminished.
Not outsourced.

Revision is not a stage to be rushed past.

It is where writing becomes real.

Revision as Ethical Governance

Revision is not merely the refinement of language. It is the point at which writing becomes accountable. It is where a writer confronts not only what a sentence says, but what it implies, what it risks, and what it may cause once released into a shared world.

This is why revision has always been the ethical center of authorship. A writer revises because they anticipate being read, misunderstood, challenged, or taken too literally. They revise because they recognize that words do not remain contained within intention. They move outward into contexts the writer cannot control but for which the writer nonetheless bears responsibility.

Seen in this light, revision is not simply craft.

It is governance.

It is where a writer decides what language deserves to remain and what language must be refused—not because it is ineffective, but because it is untrue, misleading, or disproportionate to what the writer actually means.

For those who wish to inhabit that responsibility more deliberately, a few reflective questions often clarify where authorship is actually occurring:

What did I remove that was fluent but untrue?
What reader misunderstanding am I trying to prevent?
What consequence am I willing to accept for this phrasing?

These questions do not optimize writing.

They humanize it.

Chapter Seven

Responsibility

Responsibility is the part of writing that does not scale.

Language can be generated quickly. Ideas can be drafted rapidly. Structures can be proposed in seconds. But responsibility—the obligation to answer for what words do in the world—resists acceleration. It remains bound to human awareness, human judgment, and human consequence.

This is why responsibility sits at the center of authorship, whether acknowledged or not.

Every act of writing places something into circulation. Once released, words are no longer private.

They persuade, provoke, mislead, comfort, inflame, clarify, and sometimes confuse. They enter contexts the writer cannot fully control, but for which the writer remains accountable. No tool assumes that burden.

Responsibility Begins Before Publication

Responsibility does not begin when a text is published.

It begins earlier—at the moment a writer decides what to say and why to say it.

This decision is rarely neutral. Writing always frames reality. It highlights certain facts and omits others. It adopts a tone that invites trust or suspicion. It positions the reader emotionally and intellectually. These choices shape meaning long before a reader encounters the words.

AI systems do not make these decisions.

They do not ask whether something should be written. They do not weigh relevance against harm. They do not choose silence when silence is warranted. They generate language when prompted and stop when directed.

The responsibility for initiating the act remains human.

The Burden of Meaning

To write is to assert meaning.

Even when a writer claims neutrality, the structure of the text betrays priorities. Order implies causation. Emphasis implies value. Language carries connotations beyond literal definition. Writers cannot escape this. They can only acknowledge it or ignore it.

Responsibility arises from this inevitability.

A writer who authors a claim must be prepared to explain it, defend it, or revise it. They must accept that readers will act on what is written. They must consider how the work participates in a broader discourse, whether intentionally or not.

AI cannot do this work.

It cannot recognize that meaning extends beyond correctness. It cannot perceive social consequence. It cannot feel the tension between saying what is true and saying what is wise.

Responsibility lives in that tension.

Accountability Requires a Person.

Accountability is not an abstract principle.

It requires a subject.

Someone must be answerable when words cause harm, reachable when clarification is needed, and capable of regret, correction, or apology.

Tools cannot occupy this role.

When writing is treated as a product rather than an act, accountability becomes diffuse. Blame shifts.

Responsibility dissolves into systems, platforms, or processes. This diffusion erodes trust because readers sense—often intuitively—when no one stands fully behind the words they encounter.

Authorship restores that trust by anchoring responsibility in a person.

This anchoring becomes more important, not less, when tools mediate the writing process.

The Illusion of Distributed Responsibility

One of the quiet dangers of AI-assisted writing is the illusion that responsibility can be distributed—or deferred.

When multiple tools contribute to a text, when suggestions emerge from opaque systems, when output appears without visible struggle, it becomes tempting to treat the result as collective or impersonal. If something goes wrong, blame can be displaced: the system suggested it; the data contained it; the model produced it.

This temptation must be resisted.

Assistance does not dilute responsibility.

It concentrates it.

The more a writer relies on tools, the more essential it becomes to exercise judgment. The presence of assistance does not excuse oversight; it increases the need for it. Responsibility cannot be outsourced to a system that cannot bear it.

Responsibility and Authority

Writing confers authority, even when the writer does not intend it.

Readers assume that someone who publishes has thought carefully about what they are saying—that claims have been considered and harm has been weighed. This assumption underlies the social function of writing.

AI complicates this assumption because it obscures the labor behind the text. When words appear easily, authority can feel unearned. When authorship is unclear, trust weakens.

The response to this problem is not concealment.

It is responsibility made visible through care, consistency, and revision.

Authority is not granted by effort alone.

It is earned through accountability over time.

Responsibility in Error

No writing is immune to error. Writers misunderstand facts, misjudge tone, or overlook implications. What distinguishes responsible authorship is not the absence of error, but the response to it.

Responsible writers revise when corrected. They acknowledge mistakes. They refine arguments. They adapt to new information. These acts demonstrate commitment not to ego, but to truth.

AI systems cannot do this meaningfully.

They can regenerate corrected text, but they cannot acknowledge error as error. They do not experience obligation to revise. They respond only when prompted and without memory of consequence.

Responsibility requires memory.

It requires continuity of self across time.

The Ethics of Silence

Responsibility also includes knowing when not to write.

Not every idea deserves articulation. Not every argument benefits from amplification. Not every truth needs to be spoken in every context. Writers learn this through experience, restraint, and often regret.

Silence is a form of authorship.

AI systems do not choose silence. They respond when asked. They do not feel the ethical weight of amplification. They do not recognize when adding language increases harm rather than clarity.

The decision to withhold—to revise by removal rather than addition—is one of the most human acts in writing.

It is also one of the least visible.

Responsibility and Power

Writing participates in power structures.

It shapes narratives. It influences policy. It frames identity. Writers cannot avoid this participation simply by claiming neutrality or delegating production to tools. Words move within systems of influence regardless of how they are produced.

Responsible authorship requires awareness of this fact.

It requires writers to ask how their work reinforces or challenges existing power dynamics. It requires attention to who is represented, who is excluded, and who bears the cost of misrepresentation.

AI does not perform this analysis.

It reflects patterns already present.

Responsibility lies with the human who decides whether to reproduce those patterns or resist them.

Responsibility in Teaching and Publishing

In educational and publishing contexts, responsibility extends beyond individual texts.

Educators bear responsibility for how writing is taught and assessed. Publishers bear responsibility for what they amplify and legitimize. Both must adapt to tools without surrendering standards.

This adaptation does not require omniscience.

It requires humility.

Institutions must accept that no system can perfectly police authorship. They must shift emphasis toward process, explanation, and engagement. They must recognize that responsibility is cultivated through practice, not enforced through suspicion.

AI reveals where systems have relied too heavily on artifacts rather than understanding.

It offers an opportunity—not to abdicate responsibility—but to reclaim it more deliberately.

Responsibility as the Writer's Anchor

For individual writers, responsibility offers grounding in an unsettled landscape.

When debates become polarized, when standards feel unstable, when tools evolve faster than norms, responsibility remains a constant. Writers who orient themselves toward consequence rather than convenience retain clarity about their role.

They ask:

what am I trying to say
why does it matter
who might be affected
am I willing to stand behind this

These questions do not disappear when tools are introduced.

They become more urgent.

The Weight That Cannot Be Shared

Responsibility is heavy because it cannot be shared evenly.

A tool may assist. An editor may advise. A system may distribute. But the author—the one whose name appears on the work—bears the final weight. This is not a burden to be avoided. It is the condition of authorship.

Those who seek to escape responsibility by hiding behind tools misunderstand both writing and ethics. Tools do not shield writers from consequence. They clarify where consequence ultimately rests.

Responsibility as the Final Distinction

When debates about AI and writing become abstract, responsibility offers a concrete anchor.

A system can generate language.
A system can mimic style.
A system can suggest structure.

But it cannot answer for meaning.

That task belongs to the writer.

As long as writers accept responsibility—for intent, for revision, for consequence—authorship remains intact. It is not threatened by tools. It is revealed by how those tools are governed.

Responsibility is not an accessory to writing.

It is its final distinction.

Chapter Eight

The Choice

Every writer eventually arrives at the same place. Not a conclusion, but a decision.

After the arguments have been considered, after the fears have been named, after the tools have been examined and the responsibilities clarified, what remains is not a rulebook. It is a choice—quiet, personal, and unavoidable.

Whether to use artificial intelligence in one's writing is not a question that can be settled universally. It does not submit to blanket permission or prohibition. It cannot be resolved by policy alone. It is resolved, instead, by intention.

The choice is not whether AI exists in the writing landscape.

It does.

The choice is how a writer positions themselves in relation to it.

Choice Is Not Neutral

It is tempting to imagine that refusing tools preserves purity, or that adopting them signals progress. Both positions mistake choice for identity.

Using AI does not make a writer lesser.
Avoiding AI does not make a writer virtuous.

What matters is not the presence of tools, but the reasoning behind their use. Writers who act without reflection—whether in rejection or adoption—cede agency just as surely as those who abdicate judgment to automation.

Choice, to matter, must be deliberate.

The False Comfort of Rules

In unsettled moments, writers often seek rules.

Rules promise safety. They relieve the burden of judgment. They allow responsibility to be transferred outward—to institutions, platforms, or norms. If the rules are followed, one feels protected.

But writing has never thrived under rigid rules alone.

Rules can guide practice, but they cannot replace discernment. They cannot anticipate context. They cannot account for purpose. And they cannot absolve writers of responsibility for meaning.

AI exposes this limitation sharply. No rule can specify how much assistance is too much without reference to intent, revision, and consequence. No policy can decide authorship without understanding process.

The desire for rules is understandable.

But the work remains human.

Choice as Alignment

The most useful way to approach AI is not to ask, *Is this allowed?* but *Is this aligned?*

Aligned with what you are trying to say

Aligned with how you understand authorship

Aligned with your willingness to revise and stand behind the work

Aligned with the audience you are addressing and the trust you hope to maintain

When writers ask these questions, the role of AI clarifies naturally.

For some, AI will be a brainstorming partner—useful for exploring ideas, surfacing alternatives, or breaking inertia. For others, it will serve as a drafting accelerator, a way to externalize early thinking quickly so it can be examined and reshaped. For still others, it will remain peripheral or absent, either by preference or principle.

None of these choices are inherently superior.

They are meaningful only insofar as they are intentional.

The Choice to Govern

The central choice facing writers is not whether to use AI, but whether to govern its use.

Governance means deciding where the tool is invited and where it is excluded. It means setting boundaries not because they are enforced externally, but because they preserve clarity of authorship internally.

A governed process might include:

refusing to publish unexamined output
revising until the work reflects lived understanding
rejecting language that feels persuasive but untrue
accepting responsibility for every sentence that remains

These acts do not depend on the tool.

They depend on the writer.

The Choice to Remain Visible

Another dimension of choice concerns visibility.

As AI-assisted writing becomes more common, some writers will be tempted to conceal their processes, fearing misinterpretation or judgment. Others may over-disclose, attempting to preempt suspicion through excessive transparency.

Neither extreme is required.

What matters is not disclosure as performance, but integrity as posture. Writers who are willing to explain their choices—to discuss how they arrived at a piece of work, to revise when challenged, to engage in good faith—signal authorship through behavior rather than declaration.

Visibility is not about exposure.

It is about accountability.

Choosing Restraint

One of the most powerful choices a writer can make is restraint.

AI makes generation easy. It invites expansion, elaboration, and speed. But not every possibility deserves realization. Not every articulation strengthens meaning. Restraint—knowing when to stop, when to cut, when to remain silent—is more valuable, not less, when tools remove friction.

Restraint is a choice.

It cannot be automated.

The Choice to Trust Judgment

Much of the anxiety surrounding AI arises from a lack of trust—in systems, in institutions, and often in oneself.

Writers wonder whether their judgment will be questioned, whether their work will be misread, whether their choices will be scrutinized unfairly. These concerns are real. But they cannot be resolved by abandoning judgment altogether.

The answer to mistrust is not automation.

It is discernment exercised openly and consistently.

Writers who trust their own judgment—who revise carefully, explain willingly, and accept responsibility—do not need permission to write. They have already accepted the burden that permission is meant to regulate.

Choice as Practice, Not Declaration

The choice this chapter describes is not a one-time decision.

It is a practice.

Writers will choose differently across projects, contexts, and moments. What feels appropriate in one situation may feel misaligned in another. This fluidity is not weakness.

It is responsiveness.

What matters is not consistency of method, but consistency of responsibility.

The choice to remain accountable—to govern tools rather than surrender to them—must be renewed continually.

The Freedom of Choice

There is, finally, a quiet freedom in recognizing that no tool can decide authorship for you.

No detector can take it away
No policy can grant it fully
No system can assume it on your behalf

Authorship remains a claim you make through practice.

AI does not remove that freedom.

It clarifies it.

The choice, then, is simple—but not easy.

To write with intention
To revise with care
To accept responsibility
To stand behind the work

Everything else is secondary.

Chapter Nine

The Flag

A flag is not a weapon.

It is not raised to strike, nor to conquer. It is raised to be seen—to say this is where I stand, and to make that stance visible to those who approach from different directions.

This book has not argued for or against artificial intelligence. It has argued for something older and steadier: the persistence of authorship as a human responsibility. Everything that precedes this chapter—the fears named, the tools examined, the distinctions drawn—has led here, not to a conclusion, but to a declaration.

Not everyone will agree with it.

Agreement is not required.

Clarity is.

What This Flag Is—and Is Not

The flag planted here does not claim moral superiority. It does not divide writers into camps of purity and compromise. It does not presume to speak for all practices, all contexts, or all futures.

It does something simpler.

It asserts that writing remains human work when humans remain accountable for it.

That assertion carries consequences. It rejects the idea that authorship can be inferred mechanically. It refuses the premise that tools determine responsibility. And it declines to participate in the false drama of replacement and obsolescence.

This is not optimism.

It is proportion.

The Position, Stated Plainly

Here is the position this book takes, without embellishment:

Artificial intelligence can assist writing
It cannot assume authorship

Authorship resides in agency, judgment, revision, and responsibility. These elements do not vanish when tools are introduced.

They intensify.

Writers who govern tools rather than surrender to them remain authors. Writers who accept responsibility for meaning remain accountable. Those who revise with care and stand behind their words retain legitimacy. Not because a system verifies them, but because their practice does.

This position does not deny complexity.

It refuses confusion.

Why the Flag Matters

Some will ask why such a declaration is necessary at all. Why insist on authorship when writing continues, readers continue to read, and tools continue to evolve?

Because moments of transition invite drift.

When familiar markers dissolve, when standards feel unstable, and when authority is contested, it becomes tempting to let definitions slide. Authorship becomes vague. Responsibility disperses. Judgment is replaced by metrics. In such moments, silence is not neutrality.

It is acquiescence.

Planting a flag does not end debate.

It anchors it.

It provides a reference point—a place from which disagreement can proceed without dissolving into accusation or panic.

What This Flag Rejects

It rejects the idea that writing is primarily about production.

It rejects the belief that effort alone confers legitimacy.

It rejects the fantasy that authorship can be automated or verified through surface inspection.

It rejects the framing of AI as an adversary to be defeated or a savior to be embraced uncritically. Most of all, it rejects the abdication of judgment.

No tool absolves a writer of responsibility.

No system can stand in for intent.

No detector can substitute for trust earned through practice.

What This Flag Affirms

It affirms that writing is an act of meaning-making, not merely text generation

It affirms that revision is where authorship becomes visible

It affirms that responsibility cannot be delegated to tools

It affirms that writers retain agency. Not because technology allows it, but because authorship requires it.

It affirms that clarity, restraint, and accountability matter more than novelty or speed.

A Note to Those Who Disagree

Some readers will object that this position underestimates future capabilities. They will argue that tools will grow more autonomous, more persuasive, more indistinguishable from human output.

They may be right about capability.

They are mistaken about authorship.

Authorship is not threatened by imitation. It is threatened only when humans relinquish responsibility. No increase in fluency changes the fact that meaning must be chosen, that consequences must be borne, and that accountability must reside somewhere.

If writing becomes indistinguishable at the level of output, authorship becomes more—not less—dependent on practice.

The flag stands precisely because the future is uncertain.

A Note to Institutions

Institutions reading this book may be tempted to search for policies, procedures, or enforcement mechanisms. Those have their place.

But no policy can replace discernment
No procedure can automate trust

Institutions that wish to preserve integrity must invest in process rather than proxies. They must emphasize explanation over inspection, engagement over suspicion, and responsibility over detection.

This is not a call for leniency.

It is a call for accuracy.

A Note to Writers

For writers, the flag planted here offers neither permission nor prohibition.

It offers orientation.

You are not required to justify your process theatrically. You are not obligated to perform struggle to prove humanity. You are not diminished by assistance, nor elevated by refusal.

What you are required to do if you claim authorship, is to govern your work.

To revise with care

To choose with intention

To accept consequence

To remain answerable

If you do these things, authorship remains intact. Regardless of the tools involved.

Carrying the Flag Forward

A flag planted once does not remain standing without attention.

Authorship must be practiced continuously. Judgment must be exercised repeatedly. Responsibility must be accepted again and again, across projects and contexts.

There will be moments when the line feels unclear. When assistance feels too easy. When judgment feels insufficient. These moments are not failures.

They are invitations.

To slow down
To revise more carefully
To ask harder questions of language
To recommit to responsibility

The flag does not mark an endpoint.

It marks a stance.

The Quiet Work That Remains

This book does not ask readers to defend a position publicly, nor to enter debates with sharpened arguments. Its purpose is quieter.

It asks that writers remember what they are doing when they write.

They are not competing with machines
They are not producing artifacts for evaluation systems
They are not optimizing language for plausibility

They are shaping meaning in a shared world.

That work has never been easy
It has never been automated
And it has never been optional.

The Flag, Restated

So the flag is raised here not as a challenge, but as a reference.

Writing remains human work when humans remain responsible for it.

Everything else is tool, context, and noise.

This position will not satisfy everyone.

It does not need to.

It is sufficient that it is clear.

APPENDIX

Principles for Responsible AI-Assisted Writing

What follows is not a code of conduct, nor a compliance framework. It is a set of orienting principles drawn from the arguments of this book. They clarify what ethical authorship requires when intelligent tools participate in the writing process.

They are not rules.
They are responsibilities.

1) Authorship resides in responsibility, not generation

To author a text is to accept responsibility for its meaning and its consequences. Generation—whether human or machine—does not confer authorship. Authorship begins where judgment, revision, and accountability begin.

2) Assistance does not dilute accountability

The use of tools does not distribute responsibility away from the writer. It concentrates it. The more assistance is involved, the more essential it becomes for a human to exercise discernment and to stand behind the final version.

3) Revision is the seat of authorship

Revision is where writing becomes deliberate rather than provisional. It is the moment when a writer governs language rather than merely producing it. Without revision, there is no meaningful claim to authorship.

4) Judgment cannot be automated

No system can decide what should be said, how it should be framed, or what consequences are acceptable. These are not technical decisions. They are ethical and contextual judgments that require human awareness.

5) Tools do not confer or remove legitimacy

Writing does not become legitimate through struggle, nor illegitimate through assistance. Legitimacy arises from the integrity of the authorial process and the writer's willingness to accept responsibility for meaning.

6) Disclosure is a posture, not a performance

Transparency about process is not a ritual confession. It is an orientation toward honesty and accountability. What matters is not exhaustive disclosure, but good-faith willingness to explain one's choices when challenged.

7) Detection is not evidence of authorship

No detection system can verify who authored a text or how it was written. Inference is not proof. Probability is not accountability. Authorship is established through practice, not through forensic inspection.

8) Meaning precedes fluency

Fluent language is not inherently meaningful, truthful, or responsible. Writing must be governed by intent and consequence rather than by surface plausibility or stylistic smoothness.

9) Consequence belongs to the human

When writing causes harm, misunderstanding, or distortion, responsibility does not rest with a tool or a system. It rests with the human who chose to publish and who claimed authorship.

10) Governance defines ethical use

Ethical use of AI in writing is not defined by permission or prohibition. It is defined by whether a writer governs tools through judgment, revision, restraint, and accountability.

A Minimal Accountability Framework for Writers

This template is neither a workflow nor a compliance instrument.
It is a reflective scaffold for writers who wish to operationalize responsibility without bureaucracy.

It may be used privately or shared selectively.
Its purpose is not to prove authorship.
Its purpose is to practice it.

1) Intent Note *(2–3 sentences)*

What am I trying to say?
Why does it matter?
Who is this for?

2) Boundary Statement *(1–2 sentences)*

What assistance I invited.
What assistance I refused.
Why those boundaries matter.

3) Draft Origin Summary *(2–4 bullets)*

Where language originated. (self / AI / hybrid).
Where judgment reshaped it.
Where refusal occurred.
Where meaning was clarified through revision.

4) Revision Memo *(4 short bullets)*

What I removed.
What I reframed.
What ethical risk I noticed.
What misunderstanding I tried to prevent.

5) Accountability Claim *(1 sentence)*

"I accept responsibility for the meaning and consequences of this text."

A Closing Word to Fellow Writers

Whatever tools you choose—or choose not to use—remember this:

Writing has never been about method alone.
It has always been about attention.

You write because something presses to be said. A question, a memory, a story, a way of seeing the world that does not quite fit anywhere else until you give it language. That impulse existed long before this moment, and it will remain long after today's debates have cooled and moved on.

Do not let the noise convince you otherwise.

Whether you write slowly or quickly, by hand or by keyboard, alone or with assistance, what matters is not how the words arrive on the page, but how you tend to them once they do. That you listen closely. That you revise with care. That you remain alert to what the language is actually doing—not just what it sounds like.

Writing is not proven by struggle, nor diminished by ease.

It is proven by judgment.

The willingness to choose one sentence and let another go.

The patience to refine rather than rush.

The humility to revise when meaning falls short.

The courage to stand behind what remains.

The world does not need fewer stories.

It needs stories shaped with intention, written by people who are paying attention.

Stories that come from lived experience, reflection, and care.

However you arrive at the page, arrive honestly.

However you shape the work, shape it deliberately.

When it is finished, let it go—knowing that you have done what writers have always done: not merely produced language, but made meaning.

That work has never belonged to machines.
It has never belonged to systems.

It has always belonged to those willing to take responsibility for the words they release into the world.

And it still does.

Publisher:
R.A. Wilson

Made in the USA
Monee, IL
09 November 2026

67381877R00084